Oceans of the Future

Paul Stein

The Rosen Publishing Group, Inc.
New York

Published in 2001 by The Rosen Publishing Group, Inc.
29 East 21st Street, New York, NY 10010

First Edition

Library of Congress Cataloging-in-Publication Data

Stein, Paul, 1968–
Oceans of the future / Paul Stein. — 1st ed.
p. cm. — (The library of future weather and climate)
Includes bibliographical references and index.
ISBN 0-8239-3416-0 (lib. bdg.)
1. Ocean—atmosphere interaction—Environmental aspects. 2. El Niño Current—Environment aspects. 3. Climatic changes. I. Title. II. Series.
GC190.2 .S74 2000
551.6—dc21

00-011390

All temperatures in this book are in degrees Fahrenheit, except where specifically noted. To convert to degrees Celsius, or centigrade, use the following formula:

Celsius temperature = (5 ÷ 9) x (the temperature in Fahrenheit - 32)

Manufactured in the United States of America

Contents

Introduction

Late each autumn, after months of dry, sunny weather, residents of southern California prepare for the rainy season. Jet stream winds, which remain far to the north all summer, begin shifting southward. The jet stream is a discontinuous current of air that blows at high levels in the atmosphere, 25,000 feet or higher. It's hundreds of miles wide and thousands of feet deep, and flows generally from west to east around the earth, sometimes dipping toward the equator, sometimes bending toward the poles. Huge storm systems rolling in from the Pacific Ocean follow the path of the jet stream. As autumn transitions to winter, the jet stream dips southward across southern California more often, carrying Pacific storms with it. These storms drop much needed rainfall and mountain snow on the region. This rain and snow supply the region with most of its

A mudslide threatens homes in Millbrae, California, in the wake of nonstop, torrential downpours in February 1998.

drinking water for the coming year.

In February 1998, however, the normal winter rains turned into torrential downpours. Storm after storm rolled in from the Pacific, battering southern California with rain, wind, and waves. One such storm on February 2–3 packed hurricane-force wind gusts, toppling hundreds of trees and power lines, and even blowing the roof off of an apartment building. Dozens of homes were flooded, and roads were closed by rising rivers and mudslides. Over three feet of snow fell in the mountains. Ten- to fifteen-foot waves pounded the beaches, eroding the shoreline and sending beachfront homes toppling into the churning water. The destruction was made even worse by unusually high ocean water levels, which extended the reach of the battering waves.

By February 23, southern California was waterlogged and reeling from weeks of stormy weather as another giant Pacific storm swept inland. Torrents of rain turned hillsides into rivers of mud, carrying away homes and cars. In Laguna Beach, southeast of Los Angeles, over 300 homes were damaged by mudslides, and two people lost their lives. Rushing water loosened the earth and caused giant craters to open up below homes and roads. Along Interstate 15 in San Diego, one such sinkhole measured 550 feet long, 35 feet wide, and 65 feet deep. In Santa Maria, north of Los Angeles, a highway sinkhole swallowed a California Highway Patrol car, killing the two officers inside. High winds blew down even more trees. A large eucalyptus tree fell on a car in Claremont, just east of Los Angeles, killing two students.

By the end of the month, rainfall totals across southern California were astronomical. Many locations recorded ten to fifteen inches of rain for February alone—nearly as much rain as falls usually in an entire year. An average February sees perhaps only three to four inches.

Scientists looking for clues to the cause of the unusually violent weather over southern California in February 1998 turned their attention to the oceans. That year, El Niño, a warming of ocean water in the tropical eastern Pacific Ocean, had swelled to record intensity. El Niño normally occurs every few years and most of the time causes only a modest warming of ocean water. Every once in a while, however, El Niño becomes a monster, influencing weather systems worldwide. The year 1998 was such a year.

El Niño is a prime example of the direct link between the oceans and the atmosphere. How is it that the oceans can trigger such extreme weather? Can the atmosphere, in turn, cause changes in the oceans? How will changing weather and climate patterns predicted by many scientists for the coming decades affect the oceans of the future? In order to answer these questions, we must first examine how the oceans and atmosphere interact.

1 The Air and the Ocean

The events of February 1998 are a dramatic example of how changes in the ocean can cause changes in the atmosphere. The oceans, after all, cover over 70 percent of the earth's surface and contain 97 percent of all the water on the earth. Just how do the oceans and atmosphere interact?

The oceans and atmosphere exchange heat and moisture with one another. This happens right at the surface of the water. When two objects are in contact with one another, as the surface of the ocean is with the atmosphere, heat can pass between them. Heat is just a measure of how fast molecules are moving on average. The faster the molecular motion, the warmer the temperature. When the molecules of two objects with different temperatures touch one another, the energy of both

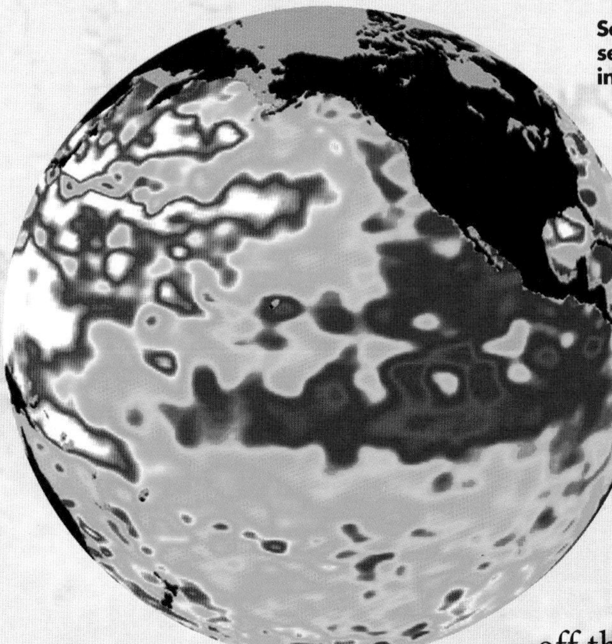

Satellite imagery helps us see that above-normal sea-surface temperatures have been developing in the Pacific and other bodies of water.

changes. The faster molecules in the warmer object slow down as some of their energy is transferred to the slower molecules of the colder object. Likewise, the slower molecules of the colder body speed up as they gain energy from bouncing off the faster molecules in the warmer object. This process, whereby objects in contact with one another exchange heat, is called conduction.

The rate at which each body warms or cools depends on the molecular properties of each object and the temperature difference between them. Whereas air temperature can rise or fall relatively quickly in response to cold or warm ocean water below, ocean temperatures rise and fall very slowly in response to warm or cold air above. This is because ocean water has a high heat capacity. It can absorb a large amount of heat from the air while warming only a relatively small amount. It takes a prolonged period of contact with warm or cold air above the ocean surface to produce a small increase or decrease in ocean temperature. Most people who have swum in the ocean or in a lake have noticed this sluggish response of water to air. It

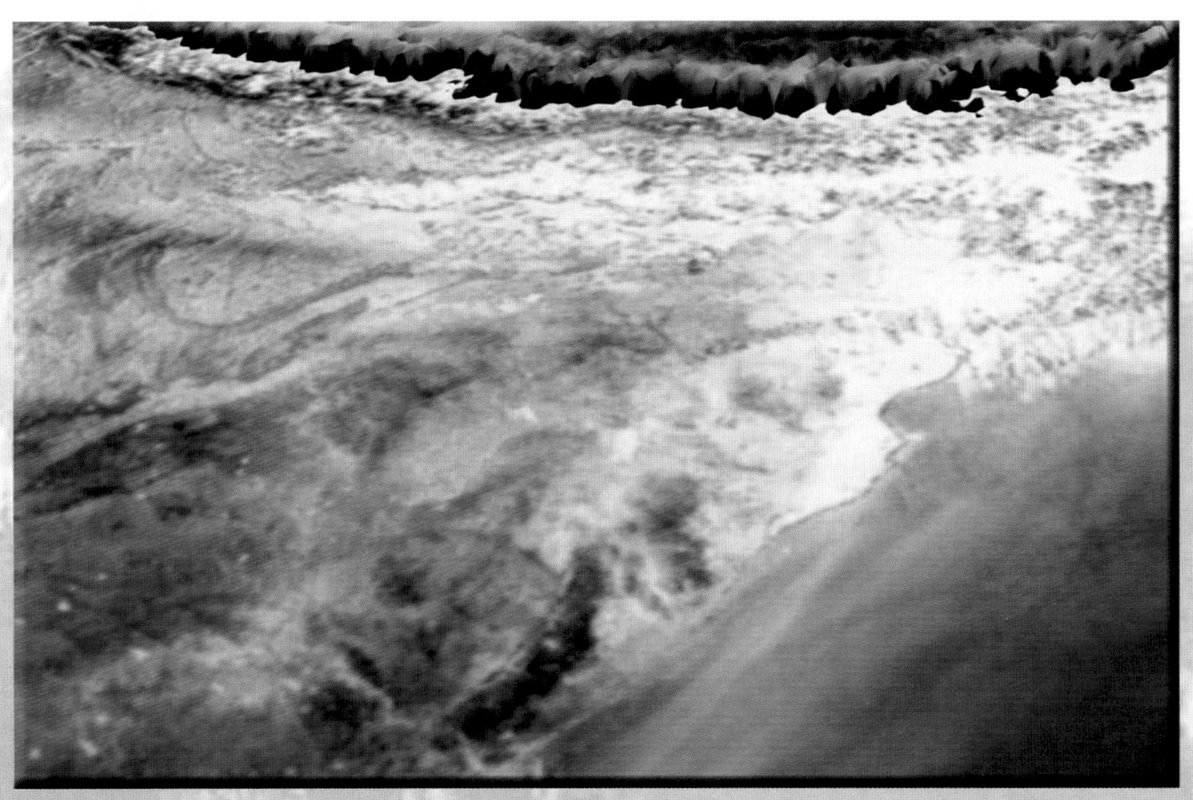
Most of the moisture that becomes rainfall is drawn from the evaporation of water in the oceans.

requires many weeks of warm air in the spring and early summer before ocean and lake water temperatures rise to a comfortable level for swimming. Even as early spring heat waves warm the air into the 80s or 90s, lake and ocean waters stay in the bone-chilling 50s and 60s.

Oceans and air also interact by exchanging water. One obvious example of this is rain, which falls from the atmosphere into the oceans or, if it falls on land, drains into rivers, which flow into the oceans. But most of the moisture that goes into producing rain in the first place comes from the oceans. Oceans add moisture to the air through the process of evaporation. Evaporation occurs as bouncing and jostling water molecules fly off into the air, becoming invisible, gaseous water vapor. The rate at which this occurs depends in large part on the temperature of the ocean. The warmer the ocean, the faster the water molecules bounce around, and the more easily they fly off and evaporate into water vapor. In the atmosphere, water vapor is what provides moisture for clouds, rain, and snow. The process whereby water changes from liquid to gas and back as it moves between the oceans, land, and atmosphere is called the hydrological cycle.

Complicating the picture is the fact that the atmosphere and oceans are in constant motion. The oceans and atmosphere interact through the force of the wind on the ocean surface. As air moves above the oceans, it pushes water along underneath. And while wind blows in many directions as a result of changing weather systems, there is a definite pattern to the way the air moves around the earth. This average global circulation of air helps drive the great currents of the oceans.

In the tropics, for example, winds blow from east to west on average. These easterly winds push against ocean water, forming ocean currents that move generally toward the west with the flow of air. As these currents encounter land masses, they turn toward the poles. As the water flows farther and farther away from the equator, it starts to come into contact with westerly winds. Winds at higher latitudes outside of the tropics generally blow from west to east. These westerly winds push against the poleward flowing ocean currents, causing them to turn more toward the east as they carry warm water northward.

Great ocean currents transport vast amounts of water all around the planet. These currents are like great rivers, many times the size of ordinary rivers. The Gulf Stream is one example of an ocean current. It originates in the warm tropical sea called the Gulf of Mexico. Flowing eastward around the southern tip of Florida, it then turns northward, parallel to the southeast coast of the United States, before heading out to sea off of Cape Hatteras, North Carolina. Moving along at up to sixty miles per day, eventually it reaches the North Atlantic Ocean and western Europe.

This is the northern edge of the Gulf Stream, which moves from the Gulf of Mexico to the North Atlantic.

The Gulf Stream does more than just move water around. Like all ocean currents, it helps distribute heat around the earth. All along the Gulf Stream, the warm water at the surface of the ocean warms the air in contact with it. The warm ocean water also evaporates into the atmosphere. The atmosphere above and near the Gulf Stream, therefore, is warmer and more humid than it would otherwise be.

The Gulf Stream is a large and long current, giving off heat and moisture into the air over a vast area. This has a major impact on

Deserts are often found near cold ocean currents. Cold currents keep adjacent warmer coastal regions dry because less cool water evaporates into the atmosphere than warm water. This reduces the amount of moisture in the air that can produce rain and snow.

climate. Europe, for example, is much milder than other parts of the world at the same latitude. One of the biggest reasons for the temperature difference is that the atmosphere over western Europe is warmed by the Gulf Stream. It's amazing to think that Paris, France, with its warm summers and relatively mild winters, is farther north than Montreal, Canada. Whereas the more northerly Paris has average January temperatures in the low 40s, Montreal shivers with an average January temperature of around 14°F. In fact, scientists estimate that Europe is an average of nine to eighteen degrees warmer than other parts of the world at the same northerly latitude because of the moderating effects of the mild ocean water.

This kind of ocean influence on climate is felt not only in Europe but all around the planet. Cold currents, such as those found off the west coasts of Africa, South America, and the United States, tend to keep adjacent coastal regions much cooler than inland areas. They also keep these regions drier. Cool water evaporates less into the atmosphere than warm water, reducing the amount of moisture in the air that can produce rain and snow. Air cooled by cold ocean water is also more dense and therefore less likely to rise and form clouds. The climate of land masses lying next to cold ocean currents, therefore, is often dry. Some of the world's great deserts, including the Atacama in Chile, the Namib in southern Africa, and the Sonoran on the Baja peninsula of Mexico, are located next to cold ocean currents.

Oceans, therefore, can significantly alter climate and weather through the exchange of heat and moisture with the air above. In fact,

The harsh storms that plagued California in 1998 gained strength when El Niño warmed ocean water in the Pacific, causing sharp temperature contrasts between northern winter air and southern tropical air.

the severe storms of February 1998 in California described earlier were fueled in part by these very same kinds of heat and moisture exchanges between ocean and atmosphere. That winter, El Niño, the warming of ocean water in the eastern tropical Pacific, grew to record proportions. The unusually warm ocean water added heat and abundant moisture to the atmosphere above it. The extra heat in the atmosphere sharpened the contrast between cold wintertime air to the north and warmer tropical air near the equator. This temperature contrast caused jet stream winds to blow faster and farther south, directing storms into southern California. These storms tapped the extra tropical moisture in the air, leading to flooding rains and mudslides.

El Niño forms as a result of shifting wind and air-pressure patterns over the length of the Pacific. Changes in the weather over the Pacific, therefore, cause a change in the ocean temperature (El Niño), which in turn causes weather patterns to shift across large parts of the world. But atmospheric changes over the Pacific that seem to start the whole process rolling are themselves directly linked to the oceans. So where does it all begin? As scientists have become increasingly aware, the atmosphere and oceans are engaged in a continuous exchange with no starting or ending point. Each affects the other in different ways and at different rates. The oceans change the weather, but the weather changes the oceans at the same time.

We've seen how the oceans and atmosphere interact. Changes in the oceans, such as the warming of water that occurs with El Niño, can significantly alter the atmosphere. Changes in the atmosphere, likewise, can have equally major effects on the earth's oceans. The oceans of the future depend in large part on the earth's atmosphere. If the earth's climate changes, so will the oceans. Next, we examine just how and why the climate of the earth may change in the coming decades.

2 Changing Climate

Between 65 and 145 million years ago, dinosaurs ruled the earth. From the spiky armored stegosaurus, to the giant plant-eating brontosaurus, to the tyrannosaurus, these ancient creatures lived in a world very different from ours. Scientists call this era the Cretaceous period. During this era, great tropical forests thrived in what is now northern Canada. Much of what would become the United States was covered by a vast, shallow inland sea—the result of much higher ocean levels than today. Most land areas were warm and humid year-round. Winter, with its frigid temperatures and snowstorms, was largely unknown. In fact, air temperatures and ocean water levels during the Cretaceous period were the highest on Earth during the last 500 million years.

Fast forward to 20,000 years ago. Great walls of ice stretched across the central part of the North American continent. A massive ice sheet covered what is now New York City, stretching all the way to the North Pole. High winds from energetic storms blew dust through the atmosphere. Dinosaurs had become extinct millions of years earlier, and what life remained was confined to a zone between the equator and the vast glaciers. With so much water locked up in ice, sea levels were hundreds of feet lower than today.

From the warm and humid Cretaceous period to the depths of the last ice age, the average temperature of the earth has varied by 15 to 30°F. Just about everyone is familiar with this kind of temperature swing. In the morning, the outside temperature is often twenty degrees cooler than in the afternoon when the Sun has warmed the land. How could such a temperature change account for such drastic swings in climate over millions of years?

The earth's climate is defined as the long-term average of weather conditions all around the globe. Weather observers record these conditions at thousands of locations worldwide, from deserts to mountains to forests, from the Arctic to the tropics, on land and at sea. Year after year this data is recorded, examined for accuracy, and averaged together. Currently, this average global temperature is around 60°F. This is five to nine degrees warmer than during the last ice age but ten to twenty degrees lower than the warmth of the Cretaceous period.

When the global average temperature changes, it does so in response to relatively slow changes in the earth's climate.

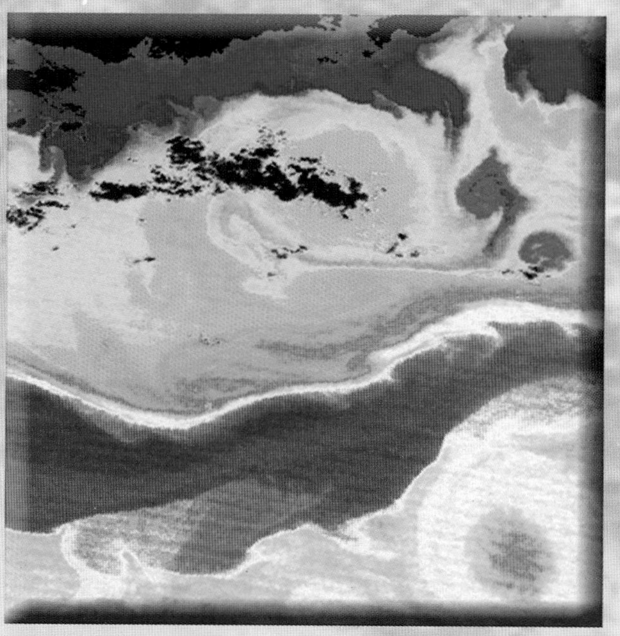

The warmer waters carried north by the Gulf Stream are depicted here in red and yellow.

Climate change is a natural part of the earth's history and occurs for a variety of reasons. One factor that causes gradual climate change over extremely long periods of time is that the orbit of the earth around the Sun slowly changes. Over periods of tens of thousands of years, the path that the earth takes around the Sun shifts slightly, along with the angle at which the earth is tilted on its axis. This alters the amount of sunlight reaching the earth and the way this sunlight is distributed between polar and tropical regions. Since the earth's temperature is driven by the warming energy of the Sun, changes in the earth's orbit can have significant changes on our climate. Scientists have linked the comings and goings of the ices ages over tens of thousands of years with these gradual orbital cycles. Other events can change climate over shorter time periods. Massive volcanic eruptions spew huge amounts of gases high into the atmosphere, where they form a haze that circles the globe. This volcanic haze can block enough

By blocking sunlight, the gases emitted by massive volcanic eruptions can lower the average global temperature by 1°F or more over the course of a few years.

incoming sunlight to lower the average global temperature by 1°F or more for a few years. Strong El Niño events, such as that which occurred in 1997–1998, can increase average global temperature by a degree or two for several years. This happens as El Niño-warmed ocean water adds heat to the atmosphere, and as cloudiness decreases over parts of the planet. Likewise, La Niña events can cool the earth by a similar amount. La Niña is the opposite of El Niño—a cooling of ocean water in the tropical eastern Pacific.

In recent years, however, scientists have been paying increasing attention to the influence of human activity on the earth's climate—in particular, the burning of fossil fuels and the effect on the average global temperature. Fossil fuels include coal, natural gas, and oil. They're so-named because they are created deep underground over millions of years from the fossilized remains of plants. Plants, like all life, are based on carbon molecules. When we burn fossil fuels to generate energy, therefore, we release this carbon into the atmosphere in the form of carbon dioxide (CO_2), a gas. Since the late seventeenth and early eighteenth centuries, when coal was first used in large volumes to drive the Industrial Revolution, CO_2 amounts in the atmosphere have increased by 30 percent. And with the gigantic energy needs of the modern world, more and more CO_2 is pumped into the atmosphere every day. Cars and trucks that run on gasoline release CO_2 into the air through exhaust. Most power plants generate power by burning fossil fuels. Our modern society has been largely built by the combustion of ancient plants.

The burning of fossil fuels releases huge amounts of carbon dioxide (CO_2) into the atmosphere and has been linked to global warming.

Carbon dioxide has a profound effect on the earth's climate. Carbon dioxide is known as a greenhouse gas. This means that it is an efficient absorber of certain kinds of radiation. The temperature of an object depends on the amount of radiation it absorbs compared with the amount it gives off. If an object absorbs more radiation than it receives, it warms. If it gives off more radiation than it absorbs, it cools.

Carbon dioxide and other greenhouse gases, such as methane, water vapor and chlorofluorocarbons, warm the atmosphere by absorbing radiation given off by the earth. The more of these greenhouse gases in the air, the more energy is absorbed and the warmer the atmosphere becomes. Scientists estimate that the average temperature of the earth was about 1°F warmer at the end of the twentieth century than at the beginning. Though one degree seems like a small variation, it's significant considering that the average planetary temperature has increased by only 5 to 9°F since the last ice age. Furthermore, the rate of temperature rise is accelerating. The warmest years of the twentieth century all occurred in the 1980s and 1990s. The 1990s were not only the warmest decade of the twentieth century, but scientists think that they were probably the warmest years of the last 1,000 years.

And while there is much debate about the cause of the recent global warming, many scientists think that some of it—perhaps most of it—is due to the increase of greenhouse gases in the atmosphere. By looking at current trends, including the rate of increase of CO_2 in the atmosphere, and using this data to run extremely complex simulations of the atmosphere and oceans, scientists think that the earth could warm by 3 to 11°F over the next century.

The atmosphere and oceans are directly linked. Changes in one result in changes in the other. Among the many possible effects of global warming, including drought, flood, and changing weather patterns, scientists are predicting significant changes in the oceans. Scientists are confident that one of the most significant effects of a warming planet will be a rise in ocean water levels. We've seen that ocean levels during the Cretaceous period and the last ice age were drastically different from those of today. How does the ocean rise and fall with changing climate? How might global warming affect ocean levels?

3 Rising Oceans

Cape Hatteras, North Carolina, juts out into the Atlantic Ocean along the eastern seaboard of the United States midway between Florida and Maine. Long, thin islands of sand, grass, scrub, and small trees protect the coast from the relentless battering of waves and tides. Known as the Outer Banks, these islands are a favorite destination for tourists. Hotels and summer houses line the shoreline, perched on the narrow strip of sand less than ten feet above sea level.

Standing guard over the fragile islands and nearby ocean is the Cape Hatteras lighthouse, the tallest lighthouse in the nation, rising over 200 feet above the dunes. The Outer Banks have become notorious as a graveyard for ships. Many ships have run aground in shallow water

The Cape Hatteras lighthouse in North Carolina had to be moved to rescue it from the eroded, unstable sands upon which it stood.

on shifting sandbars, or have been wrecked in the great storms that sweep the coastline. For 130 years, ship captains have looked for the guiding beam of the lighthouse in times of dangerous weather.

By the 1990s, however, it was the Cape Hatteras lighthouse that needed rescuing. Years of storms and beach erosion had swept away the thin strip of sand that lay between the massive structure and the churning surf of the Atlantic Ocean. The historic lighthouse was in danger of being undermined and toppled by the very waters over which it stood guard. After years of planning, and at a cost of $12 million, engineers took action. They lifted the entire structure—all 2,800 tons of it—using special, high-powered jacks. Then, they slowly set it on steel beams that rested on rollers. Moving at a snail's pace, the great old lighthouse was nudged safely away from the water's edge and set down 2,900 feet from its original position.

The Cape Hatteras lighthouse serves as a dramatic example of the fragile nature of our coastal regions, in particular the barrier islands that stretch along so much of the southern and eastern coast of the

United States. Currents and tides carry sand to and fro, creating sandbars and adding to islands here, eating away at beaches there. Storms hammer the coastline with waves, causing further beach erosion. This relentless action of ocean water poses a constant threat to all coastal development, even 130-year-old lighthouses. If global warming continues through the coming decades, scientists are predicting that ocean levels will rise. Rising oceans will enhance the destructive nature of waves, currents, and tides, and may flood thousands of square miles of low-lying land along the world's coastlines. Many thousands of homes, hotels, and buildings constructed along the seashore in recent decades will be in jeopardy. With rising sea levels, what will be the fate of oceanside communities in the coming years? And how does the level of the oceans rise and fall?

Daily tides and storms eroded and sculpted these rocks at Canon Beach, Oregon.

We're all familiar with the normal rise and fall of ocean tides. Ocean water is pulled back and forth across the earth by the gravitational tug of the Moon and the Sun. In some parts of the world, the difference between high and low tide can be over twenty feet. But ocean water

Sea levels do not rise when ocean-based polar icebergs melt, since water added to the ocean equals the amount of water that was displaced by the iceberg.

levels can also change over much longer time periods. One way that ocean levels can rise is through a process known as thermal expansion. This process occurs as ocean water warms. The warmer the water, the faster the water molecules move around and bounce off one another. This jostling and bouncing creates more space between the molecules. Warmer ocean water therefore expands, causing the surface of the oceans to rise. The amount of rise depends on the temperature of the water: The warmer the water, the more it will expand.

Another way that global warming may cause sea levels to rise is through the melting of glaciers. With a predicted two to six degree rise in average global temperatures, computer simulations estimate that between 33 and 50 percent of mountain glaciers may melt away. Meltwater from glacial ice flows into the oceans, causing sea levels to increase. Likewise, during a colder climate such as that which existed 20,000 years ago, glaciers grow larger as more snow falls and compacts into ice. Most of the water that goes into making the snow is evaporated from the oceans. During great ice ages, therefore, sea levels plunge as ocean water evaporates and is converted into glacier-building snow.

Polar ice also would melt in a warmer world. However, not all melting polar ice would contribute to rising sea levels. Much of the ice around the poles, especially the North Pole, is already floating on ocean water. These massive ice sheets and icebergs weigh down on the ocean water, displacing it and forcing it upward elsewhere. If this floating ice melts, the amount of meltwater added to the ocean would equal the amount of ocean water displaced by the ice. The result would be no change in sea level. However, the situation is different with land-based polar ice caps, such as those over Greenland and parts of Antarctica. These ice sheets exert no displacing force on ocean water levels. Meltwater from these land-based ice sheets, therefore, would raise ocean levels. But scientists aren't sure how much of these vast ice sheets would melt in a warmer world. If the air warms over Greenland and Antarctica, it would still remain below freezing much of the time. Warmer global temperatures would cause more ocean water to evaporate, adding moisture to the air. Moister air, in turn, would contribute to heavier snowfall over Greenland and Antarctica. Snow helps ice caps to grow by compacting into ice year after year. If enough snow falls in a warmer world,

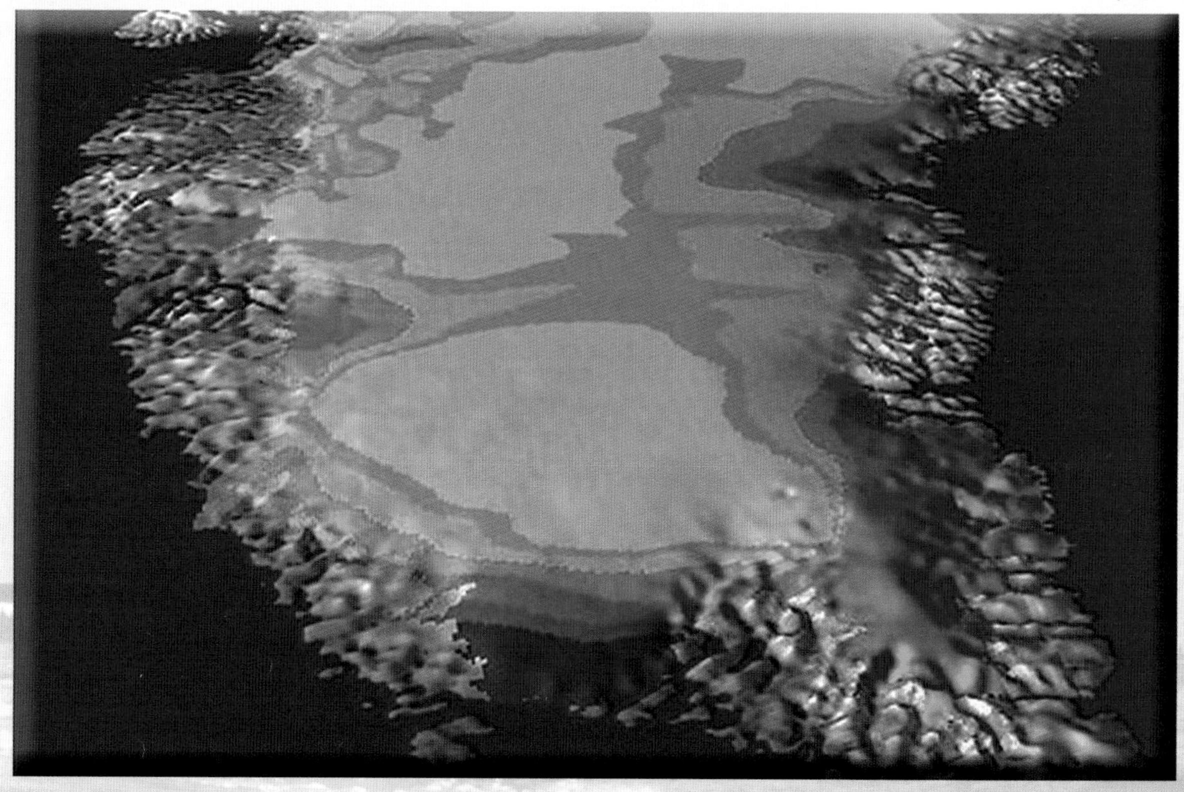

Research conducted by NASA indicates the shrinkage of the Greenland ice cap—more than three feet per year—which could contribute to rising sea levels over time.

the ice could grow faster than it melts. However, research published in the summer of 2000 by NASA indicates that the Greenland ice cap is shrinking around its edges at the rate of around three feet per year.

The rate at which ocean levels rise or fall with changing climate depends on the movement of continents. In particular, some sections of land slowly rise over time, while others slowly fall. This happens when people remove freshwater from the ground. As cities grow, more and more water is being removed from the earth. This causes ground levels to sink. Sinking land, combined with rising oceans, would lead to severe flooding along coastlines as water spreads inland. Land levels also change as the result of ice ages. The great ice sheets that spread across continents weighed heavily on the earth, pressing it down. When the ice retreated and the weight was removed, the depressed ground slowly rose back up over thousands of years. Even though we can't see it happening, some sections of North America continue to slowly rise after the last ice age, 20,000 years ago.

Scientists estimate that over the last century, ocean levels around the world rose anywhere from four to ten inches. The variation depended on whether local land areas rose or fell. Of the four to ten inch rise, three-quarters of an inch to two inches were due to the melting of glaciers and ice caps in our slowly warming climate. Based on current trends, and on the predicted three to eleven degrees of additional warming in the next 100 years, scientists think ocean levels could rise by anywhere from 5.5 inches to as much as three feet. What might be the consequences of higher sea levels in the future?

An estimated 50 percent of the world's population lives within sixty miles of an ocean coastline. The seashore attracts many vacationers as well as permanent residents. New homes, hotels, and other buildings have been built right along beaches or coastal islands like the Outer Banks of Cape Hatteras. Many of these structures may fall victim to rising seas by 2100, especially as higher water lends more destructive potential to storm-driven waves. Scientists estimate that a twenty-inch rise in ocean levels through the next century would flood 3,300 to 7,300 square miles of land across North America.

In addition to threatening human-built structures, rising oceans would endanger the unique plant and animal life of coastal zones. These coastal ecosystems often consist of saltwater marshes

Bangladesh, where about 36 million people live in low-lying areas, is especially vulnerable to flooding from storm surges (large masses of water). This flooding will worsen as sea levels rise.

or swamps, with their unique blend of birds, fish, crabs, and mollusks. Over two-thirds of edible fish depend on coastal marshes and swamps. Already, coastal wetlands are being lost at a rate of 0.5 to 1.5 percent each year. Researchers think that rising oceans may eliminate 50 percent of existing coastal wetlands across North America by 2100.

Certain coastlines are more vulnerable than others. Countries with large sections of low-lying land, only a few feet above sea level, will be most likely to flood. One such country is Bangladesh, tucked under the eastern wing of India on the Bay of Bengal. Bangladesh has a population of 120 million, 30 million of whom live on land that lies less than ten feet above sea level, and 6 million of whom inhabit land only three feet or less above sea level. To make matters worse, this part of the world also experiences violent tropical cyclones. These storms sweep inland from the oceans, flooding low-lying areas and causing terrible devastation. One such cyclone in 1991 killed over 100,000 people. In 1970, another storm killed over 250,000.

Most of the fatalities from these tropical cyclones occur in the storm surge, a large mass of water pushed inland by the storm's high winds. Storm surges can cause ocean levels to rise by twenty feet or more. Bangladesh, with 25 percent of its land area at less than six feet above sea level, is especially vulnerable to storm surges. Rising ocean levels, combined with sinking land in that part of the world, will allow cyclones and storm surges to spread devastation over much greater areas than ever before. Worldwide, scientists think that the number of people vulnerable to storm surges, currently around 46 million, will increase to 92 to 118 million by 2100 as ocean levels rise.

In the United States, heavily populated coastal areas such as New York City and southern California will also be threatened by rising water and storm surges. In New York, some computer simulations predict that water levels may rise over thirty inches by the end of the twenty-first century. While this won't be enough to flood the city, it will cause serious flooding when combined with storm-driven waves. The New York City subway system is especially risk-prone. A storm surge of ten feet, such as might occur with a strong hurricane or severe winter storm, would cause water to pour into the city's subway tunnels, paralyzing this critical mode of transportation. Other low-lying areas would be swamped, such as the regional airports and waterfront sections of lower Manhattan. On the other side of the country, in southern California, more and more beachfront homes will wash away with rising sea levels. In the El Niño event of 1997–1998, warm ocean water along the California coast rose as a result of thermal expansion. When the powerful Pacific storms of February came rolling ashore, high water worsened the already destructive effects of pounding waves. Many oceanside homes were damaged or destroyed. This type of scenario may become more common in the coming decades if ocean levels rise.

Some of the locations most threatened by rising sea levels are islands—in particular, the small islands that dot the tropical seas worldwide. Many of these islands are built on ancient coral reefs and are less than ten feet above sea level. Examples include islands in the Bahamas, and the Maldives and the Seychelles in the Indian Ocean. Since these islands are so tiny and so close to sea level, rising waters would leave little high ground for residents to move to. Furthermore, rising salty ocean water may contaminate fresh drinking water supplies by seeping into underground reservoirs. Many of these small island countries depend on tourism for their economies. As rising ocean levels flood beaches and contaminate drinking water, local economies may be devastated.

Harsh weather often devastates coastal areas.

Rising ocean levels are one of the most destructive of the potential effects of global warming. It's also one of the outcomes of global warming about which scientists are most certain. There's little doubt that rising temperatures will lead to rising ocean levels. The only questions are how much rise there will be and how fast it will occur. But the oceans will also react to a warmer world in other ways. In the next chapter, we will examine other possible consequences that global warming may have on the oceans of the future.

4 Warmer Waters

A summer day at the beach can be exciting. Most beachgoers look forward to warm ocean water—the warmer the better. Water temperatures in the 70s or 80s are considered ideal, while readings in the 50s and 60s are considered bone-chilling. From the point of view of the average beach visitor, warmer oceans of the future won't be such a bad thing. However, one of the possible outcomes of warmer oceans—higher ocean levels—may create problems for the beachgoer. There are several other changes in the oceans that global warming may bring about. Warmer oceans will certainly affect the plants and animals that live under the waves. Scientists are particularly worried about coral. Coral lives in

Rising ocean temperatures will threaten coral reefs.

shallow, warm water across the tropical seas. Coral reefs support a myriad of ocean-dwelling animals and plants, and help protect coastlines from the pounding of ocean swells. As water levels rise, coral will be covered by deeper water. The deeper the water, the cooler the water becomes. Luckily, most coral will be able to grow and keep pace with rising ocean levels.

Of greater concern will be rising ocean temperatures. Warmer water may damage or destroy many coral reefs through the process of bleaching. Coral contains a microscopic algae called zooxanthellae. This algae is what gives coral its wide spectrum of colors and also provides food and oxygen that help the coral grow. Coral thrives in ocean water temperatures of 77 to 84°F. When ocean waters reach 89.6°F for a prolonged period of time, the coral expels its color-giving algae and turns white, or bleaches. Without the algae to provide nutrients, the coral dies. While some species of coral may adapt to warmer oceans, others will not. Already, marine biologists have observed widespread bleaching of coral reefs in Florida, the Indian Ocean, and Australia. With water ocean temperatures likely to continue rising in the decades to come, many of the world's coral reefs will be endangered.

Scientists are less sure about the effects warmer oceans will have on fish. Certain species of fish live in distinct zones of water temperature. Unlike coral, which is fixed to a certain spot, fish can migrate to keep pace with changing ocean temperatures. Commercial fishing areas may shift as fish that have thrived in certain areas for years are forced away. This may hurt fishing industries in some countries but provide new opportunities for fishing in others. El Niño illustrates the

Worldwide changes in ocean temperature could drastically affect fishing cultures in many nations. Migration of fish due to these changes could shift commercial fishing to different areas.

relationship between changing ocean temperatures and a regional fishing industry. Normally, cold, nutrient-rich water off the northern Pacific coast of South America supports abundant schools of fish. When El Niño–warmed waters spread into this region, fish became scarce and the regional fishing industry, especially that of Peru, suffers. And what of El Niño in a world with warmer oceans? Since El Niño is a periodic warming of ocean water, will El Niño become more intense and occur more frequently? We've already seen how El Niño can affect weather patterns thousands of miles away. In addition to storms across the southern United States, including the ones that brought flooding and mudslides to California in February 1998, El Niño also causes a variety of weather pattern changes in other parts of the world. El Niño is known to bring drought conditions to northern Australia and Indonesia, while drenching normally dry parts of northern South America with flooding rains. In Africa, El Niño has been known to bring devastating drought

El Niño has been known to bring devastating droughts to the Sahel, a dry grassland just south of the Sahara Desert.

to a region known as the Sahel, a dry grassland that stretches from west to east across the continent just south of the Sahara Desert. Hurricanes in the Atlantic Ocean are much reduced in number and intensity during El Niño years because of strong winds in upper levels of the atmosphere. These jet stream winds are caused by the intensified contrast between El Niño–warmed air temperatures to the south and colder air to the north. As the winds blow across the Atlantic Ocean, they rip apart developing hurricanes.

El Niño alternates with La Niña, a cooling of Pacific Ocean water temperatures. Many of the effects of La Niña are the opposite of El Niño: dry weather in Peru and Ecuador, above-average rainfall in Australia and Indonesia, and increased hurricane activity in the Atlantic Ocean. Scientists do not yet completely understand the processes that lead to the development and decay of El Niño and La Niña. Therefore, they aren't sure how these periodic changes in regional ocean temperatures would change if oceans warmed globally. In recent decades, there is some evidence that the number of El Niño events is outpacing the number of La Niña events. From 1950 to 1980, for example, there were about an equal number of La Niña and El Niño events. Since 1980, scientists estimate that El Niño has occurred an average of twice as often as La Niña. The two strongest El Niño events of the century have both occurred since 1980. The longest El Niño on record occurred from 1990 to 1995. Since El Niño is a warming of ocean waters, it seems logical that it would occur more frequently and with greater intensity as ocean and air temperatures rise globally. However, scientists are very cautious on this

Scientists still debate whether the intertropical convergence zone, a place of unsettled weather, will spawn more or fewer hurricanes as the earth gets progressively warmer.

issue. In addition to an incomplete picture of the mechanisms that cause El Niño and La Niña, another problem is a lack of historical data on these two important oceanic phenomena. Scientists just haven't been observing them long enough to tell whether recent strong and frequent El Niño events are a sign of things to come or are just a passing trend.

We've seen that El Niño and La Niña can affect hurricane development in the Atlantic Ocean. Hurricanes gain their energy from warm ocean water, generally 80°F or warmer. Does this mean that warmer oceans of the future will spawn more frequent and more intense hurricanes? Will the area affected by hurricanes expand?

Not necessarily. Hurricanes require more than warm ocean water to develop. They also need a preexisting area of showers and thunderstorms from which to form. These clusters of showers and thunderstorms are commonly found in what meteorologists call the intertropical convergence zone, or ITCZ for short. The ITCZ is a belt of clouds and unsettled weather that extends around the globe near the equator. Air tends to converge, or come together, near the equator. Wind blowing in from the northeast collides with wind blowing in from the southeast. Air piles up and rises high into the atmosphere, carrying abundant tropical water vapor with it, forming clouds, showers, and thunderstorms. In certain parts of the world, where the ITCZ extends over warm tropical waters, some of these clusters of showers and thunderstorms become better organized and begin to develop their own circulation of air. Air spirals into the center of the cluster near the surface of the ocean, picking up moisture from the ocean water. It then rises high into the atmosphere, releasing moisture in the form of additional clouds and storms. Other air continues to spiral in from below to take the place of the rising air. If the system continues to intensify, the spiraling air turns with greater speed and strength, and the storms become more widespread and intense. When winds reach seventy-four miles per hour or higher, the system becomes a hurricane.

In a warmer world, it is unlikely that the ITCZ would expand or change much from its present state. Therefore, the number of hurricanes and the location where they form would probably not change much. Some scientists even argue that the frequency of hurricanes

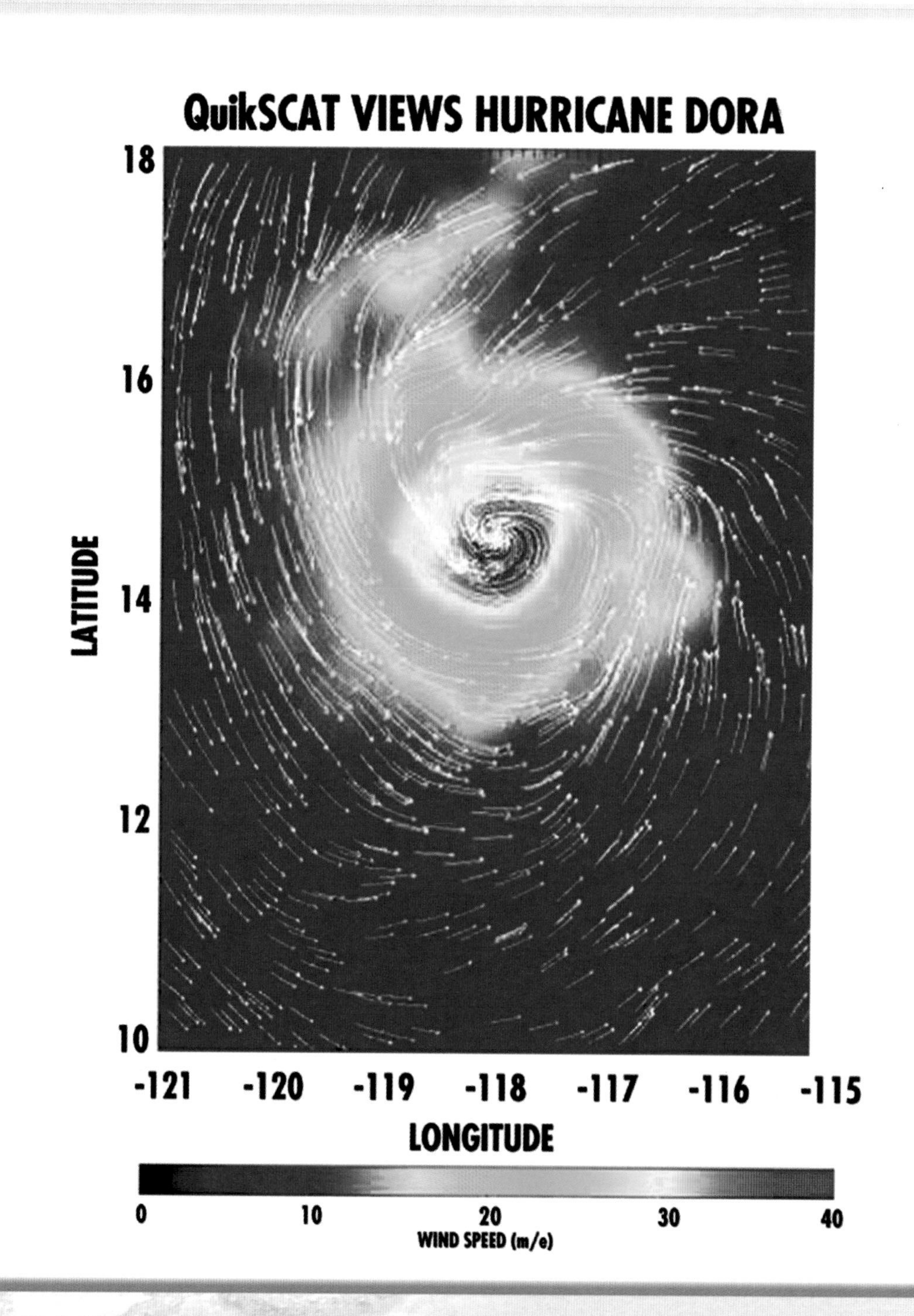

NASA's SeaWinds radar instrument captured this image of Hurricane Dora in August 1999. Scientists hope to perfect their predictions of the intensity, occurrence, and movement of storms.

might diminish slightly in the future as atmospheric temperatures rise. This is because the storms that form a hurricane grow less quickly and with less intensity if the temperatures in upper levels of the atmosphere are higher. However, other scientists do think that hurricanes of the future may become more intense than today as they spin over warmer ocean waters. Using computer simulations of the ocean and the atmosphere, one group of researchers has predicted that wind speeds in hurricanes could be 10 percent higher than today. A hurricane with 100 mph winds today, therefore, may strengthen to a hurricane with 110 mph winds by later this century.

Another outcome of warmer oceans is more subtle than giant spinning storms but potentially just as important. As we've seen, most of the moisture in the atmosphere comes from water evaporated from the surface of the oceans. Evaporation occurs as individual water molecules fly free from the ocean surface, turning into invisible, gaseous water vapor. The higher the ocean water temperature, the faster the water molecules move around and the more easily they fly off and evaporate. In a warmer world with warmer oceans, evaporation rates will increase. The more evaporation, the more moisture in the air.

Atmospheric moisture produces clouds, rain, and snow. Some scientists think that rain and snow may become heavier, on average, as evaporation moistens the air in a warmer world. But the outcome is not certain. The warmer the water, the more evaporation from the oceans, and the more humid the air becomes. The greater the humidity in the air, the more clouds form. Clouds, however, can act to lower land and

The prevalence of different types of clouds, which alternately block sunlight or absorb the earth's radiation, may affect global warming in years to come.

water temperatures by blocking incoming sunlight. Could warmer oceans, therefore, tend to suppress global warming by increasing cloud formation? The picture is further complicated by the fact that different kinds of clouds have different effects. High, thin clouds tend to produce an additional warming of the atmosphere as they let sunlight through but absorb the outgoing radiation from the earth. Low, thick clouds can cool the air as they block incoming sunlight. Which kinds of clouds will predominate in a warmer atmosphere? Scientists are not yet sure.

Warmer oceans may also trigger surprising changes in ocean currents such as the Gulf Stream. These currents help shape the earth's climate by carrying warm and cold water toward the poles and the equator. The Gulf Stream is but one part of a vast ocean circulation, called the thermohaline circulation by scientists. The name comes from *thermo*, meaning "heat," and *haline*, meaning "salt." It carries warm water northward into the North Atlantic, then sinks down into the cold depths of the ocean where it slowly travels south, crossing the equator and passing between Antarctica and the southern tip of Africa. From there it meanders around Southeast Asia and into the Pacific where it rises again to the surface south of Alaska. The surface current then retraces its steps, wandering back through the Indian Ocean and into the Atlantic where it returns as the Gulf Stream. The entire trip can take over a thousand years.

The thermohaline circulation is driven not only by the force of wind pushing against water but by changes in the saltiness, or salinity, of the oceans. The focus for changing salinity is the North Atlantic. Here, strong westerly winds evaporate ocean water. When ocean water evaporates into the air, it leaves its salt behind, causing the salt content of the ocean water to increase. Saltier water is more dense and sinks to the bottom of the sea. It's this sinking motion that is so important as a driving mechanism of the thermohaline circulation.

Some scientists speculate that global warming may disrupt the thermohaline circulation by adding more freshwater to the North Atlantic. This may happen as rain and snowfall increase over the region and as ice caps on Greenland melt, sending freshwater into the sea. More freshwater may dilute the salt content of the North Atlantic and make the ocean water less dense, interrupting the all-important sinking motion. Breaking this link in the circulation may have frightening consequences, including a chilling effect on Europe. As we've seen, Europe's climate is dependent on the warming effects of ocean circulation. If the Gulf Stream weakens, Europe may become colder, and as El Niño illustrates, changes in ocean water temperature in one part of the world have global repercussions.

The Gulf Stream is but one part of a vast ocean circulation, called thermohaline circulation by scientists.

Some scientists trace rapid worldwide cold snaps over 10,000 years ago to changes in the flow of the thermohaline circulation.

Conclusion

The oceans and the atmosphere speak in a natural dialogue, the motion of each affecting the other. The atmosphere responds swiftly to the ocean, whereas the ocean only sluggishly answers to the atmosphere. The conversation has no end and no beginning, and is spoken in a language we don't fully understand.

Storms come and go, tossing the seas into a heaving fury. While these tempests can cause great destruction, of equal or greater concern are atmospheric changes that occur much more slowly. Compared to the violence of a hurricane, global warming may seem like merely a whisper across the surface of the sea. But as oceans slowly warm in response to the warming air above, they may threaten coastal populations by rising almost imperceptibly at first. Whereas the flash floods that race down streams and rivers quickly retreat, decades of ocean flooding may bring lasting damage. The oceans of the future will probably have surprises in store for us.

Like a great blanket, the atmosphere lies above the ocean. By pumping more and more greenhouse gases into the sky, we may be warming the blanket. Underneath lies a drowsy giant. As with all giants, the oceans possess great power. If global warming continues, we may see the giant awaken and rise, like Poseidon, god of the ocean in Greek mythology.

Glossary

bleaching The process by which coral turns white and begins to die. One cause of bleaching is unusually warm ocean water.

climate The average weather conditions over a long period of time, generally decades or more.

conduction The process whereby two objects in contact with one another exchange heat energy.

Cretaceous period An ancient era lasting from around 145 million to 65 million years ago. It was a much warmer period than today and was the era during which the dinosaurs lived.

El Niño Warming of ocean water in the tropical eastern Pacific Ocean. When El Niño becomes particularly strong, it can affect weather patterns worldwide.

evaporation Process whereby liquid water changes into invisible, gaseous water vapor.

fossil fuel Any fuel made from the decayed remains of ancient plant life; includes coal, natural gas, and oil.

global warming Warming of the planet due to increasing amounts of greenhouse gases in the atmosphere.

greenhouse gases Any gases that efficiently absorb outgoing radiation from the earth. The main greenhouse gases are water vapor, carbon dioxide, methane, nitrous oxide, chlorofluorocarbons (CFCs), and ozone.

Gulf Stream A large ocean current, beginning in the Gulf of Mexico south of the U.S., then curving around the southern tip of Florida and northeastward into the Atlantic Ocean.

hydrological cycle The transfer of water in liquid, frozen, and gaseous form between the oceans, the land, and the atmosphere.

Industrial Revolution The rapid growth of factories and industry in the eighteenth and nineteenth centuries, supported through the burning of coal for energy.

intertropical convergence zone (ITCZ) The belt of clouds, showers, and thunderstorms that rings the earth near the equator. It is caused by colliding flows of air, one from the northeast and one from the southeast. Most hurricanes are born from disturbances in the ITCZ.

jet stream A fast-moving current of air high in the atmosphere, usually over 25,000 feet. The jet stream generally flows from west to east, though it can make large curves toward the north and south. It shifts toward the equator in winter and back toward the poles in summer.

La Niña A cooling of ocean water in the eastern tropical Pacific Ocean. Like El Niño, a strong La Niña can influence weather patterns worldwide.

storm surge A rise in ocean levels along the coast that occurs when storm winds push ocean waters up against the coastline.

thermal expansion The process by which water increases in volume as a result of increasing water temperature.

thermohaline circulation The global circulation of ocean water, consisting of ocean currents both at the water surface and deep in the ocean's depths. The thermohaline circulation is driven by the force of wind pushing against surface water and by changes in ocean water saltiness that force water to sink to the bottom of the sea, especially in the North Atlantic.

water vapor The invisible, gaseous form of water.

For More Information

In the United States

American Meteorological Society (AMS)
45 Beacon Street
Boston, MA 02108-3693
(617) 227-2425
Web site: http://www.ametsoc.org/AMS
The AMS is the premiere professional meteorological organization in the United States.

Climate Prediction Center (CPC)
World Weather Building
5200 Auth Road, Room 800
Camp Springs, MD 20746

(301) 763-8000
Web site: http://www.nnic.noaa.gov/cpc
The Climate Prediction Center's Web site provides information about El Niño and La Niña and current global temperature trends.

Intergovernmental Panel on Climate Change (IPCC)
c/o World Meteorological Organization
7 bis Avenue de la Paix, C.P. 2300
CH-1211 Geneva 2
Switzerland
Web site: http://www.ipcc.ch
The IPCC issues scientific reports predicting future climate change, including potential ocean level rises and consequences for coastal ecosystems.

Smithsonian Ocean Planet
National Museum of Natural History
10th Street and Constitution Avenue NW
Washington, DC 20560
Web site: http://seawifs.gsfc.nasa.gov/ocean_planet.html

Weatherwise Magazine
Heldref Publications
1319 18th Street NW
Washington, DC 20036-1802
(202) 296-6267
Web site: http://www.weatherwise.org
A popular magazine about all things weather-related. Find it at your local library or newsstand.

In Canada

OceansCanada
Station 12E239
Department of Fisheries and Oceans
200 Kent Street
Ottawa, ON K1A 0E6
(613) 990-6840
Web site: http://www.oceanscanada.com

For Further Reading

Arnold, Caroline. *El Niño: Stormy Weather for People and Wildlife*. New York: Clarion Books, 1998.

Erickson, Jon. *The Mysterious Oceans*. Blue Ridge Summit, PA: TAB Books, 1988.

Ganeri, Anita. *The Oceans Atlas*. New York: DK Publishing, 1994.

Phillips, Anne W. *The Ocean*. New York: Crestwood House, 1990.

Sayre, April Pulley. *El Niño and La Niña: Weather in the Headlines*. Brookfield, CT: Twenty-First Century Books, 2000.

Stevens, William K. *The Change in the Weather: People, Weather, and the Science of Climate.* New York: Delacorte Press, 1999.

Suplee, Curt. "El Niño/La Niña." *National Geographic*, March 1999, pp. 72–79.

Williams, Jack. *The Weather Book*. 2nd edition. New York: Vintage Books, 1997.

Index

O

P

R

S

T

V

W

About the Author

Paul Stein has a B. S. in meteorology from Pennsylvania State University. He has eight years' experience as a weather forecaster, most recently as a senior meteorologist for the Weather Channel. Currently he develops computer systems and software that display and process weather-related data.

Photo Credits

Cover image © Weatherstock: waves from ocean storm and El Niño.
Cover inset © NASA/GSFC Visualization Analysis Lab: Hurricane Floyd.
Front and back matter © Weatherstock: ocean storm from El Niño.
Introduction background © Weatherstock: flooding in Arizona.
Chapter 1 background © SeaWiFS Project, NASA/Goddard Space Flight Center (GSFC) and Orbimage: water off of Peninsula Valdes.
Chapter 2 background © SeaWiFS Project, NASA: Rio de la Plata, Argentina.
Chapter 3 background © Photo Researchers, Inc.: Cape Hatteras lighthouse.
Chapter 4 background © Corbis: Great Barrrier Reef, Australia.
Pp. 6, 16, 34, 37 © AP/Worldwide; p. 10 © TOPEX/Poseidon project, NASA/GSFC and Orbimage; p. 11 © Scientific Visualization Studio, NASA/GSFC; p. 13 © SeaWiFS Project, NASA/GSFC and Orbimage; p. 14 © Pictor; p. 21 © University of Miami using 11 and 12 micron bands, by Bob Evans, Peter Minnett, and coworkers; pp. 22, 24, 29, 30 © Photo Researchers, Inc.; pp. 28, 40, 42 © Corbis; p. 32 © NASA/GSFC Scientific Visualization Studio; p. 41 © Terje Rakke/Image Bank; p. 44 © GOES Project Science Office; p. 48 © Weatherstock.

Series Design and Layout

Geri Giordano